I Used To Be Afraid Of... Cater

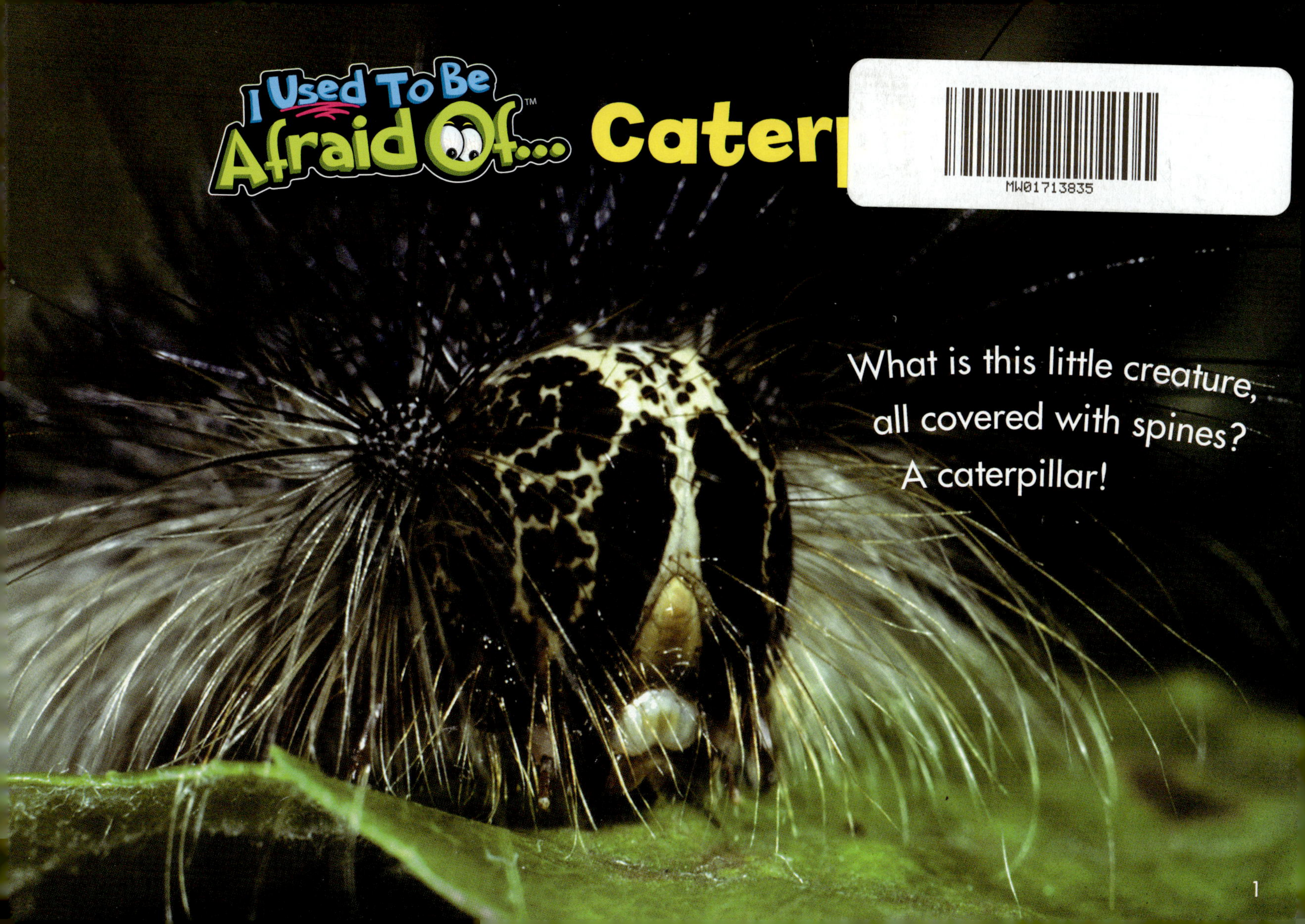

What is this little creature,
all covered with spines?
A caterpillar!

Spiny, bumpy, wriggly caterpillars creep through woods and fields and gardens. They inch up plants to eat the leaves. They even eat the vegetables we grow.

They weave their webby nests in trees. They look strange and scary. How would you feel if a caterpillar crawled on you? If you knew all about caterpillars, you would not be afraid. Caterpillars are full of great surprises.

I USED to be AFRAID of CATERPILLARS, but now I know . . .

Butterflies of all kinds were once crawly caterpillars. Moths start out as caterpillars, too. A monarch caterpillar hatches from a tiny egg. It eats and grows until it is big and fat.

Then it hooks onto a branch. It forms a hard case around its body. Now it is a pupa, or chrysalis. Inside its case, the pupa slowly changes. At last the case breaks open. A beautiful monarch butterfly crawls out.

I USED to be AFRAID of CATERPILLARS, but now I know . . .

Butterflies fly around the garden on summer days. They drink nectar, a sweet liquid they find in flowers. And they help the plants. They spread tiny grains of yellow pollen from flower to flower. This lets the plants make seeds.

Many moths fly at night. You may see them fluttering at your window.
What if there were no caterpillars? There would be no butterflies or moths.

I USED to be AFRAID of CATERPILLARS, but now I know . . .

Caterpillars are important in other ways. They are food for birds and other animals. A robin brings juicy caterpillars to the babies in her nest. The little birds peep and open their beaks wide to get it.

This wasp has caught a caterpillar. She will stuff it into her nest and leave it there with her eggs. When the eggs hatch, the young will eat the caterpillar.

I USED to be AFRAID of CATERPILLARS, but now I know . . .

A caterpillar called the silkworm makes fine strands of silk. The silkworm spins the strands into a fat cocoon. Inside its cocoon, it changes into a white moth.

People use the silk from the cocoons. They soak the cocoons in water and pull the silk strands apart. Then they weave the strands into soft, shiny silk cloth.

I USED to be AFRAID of CATERPILLARS, but now I know . . .

Caterpillars are fun to watch. The inchworm seems to measure plants as it crawls around. It moves its body one end at a time. The front moves first. Then the back catches up.

Woolly bears crawl across roads and paths on warm fall days. Some people say the more brown the woolly bear wears, the milder winter will be.

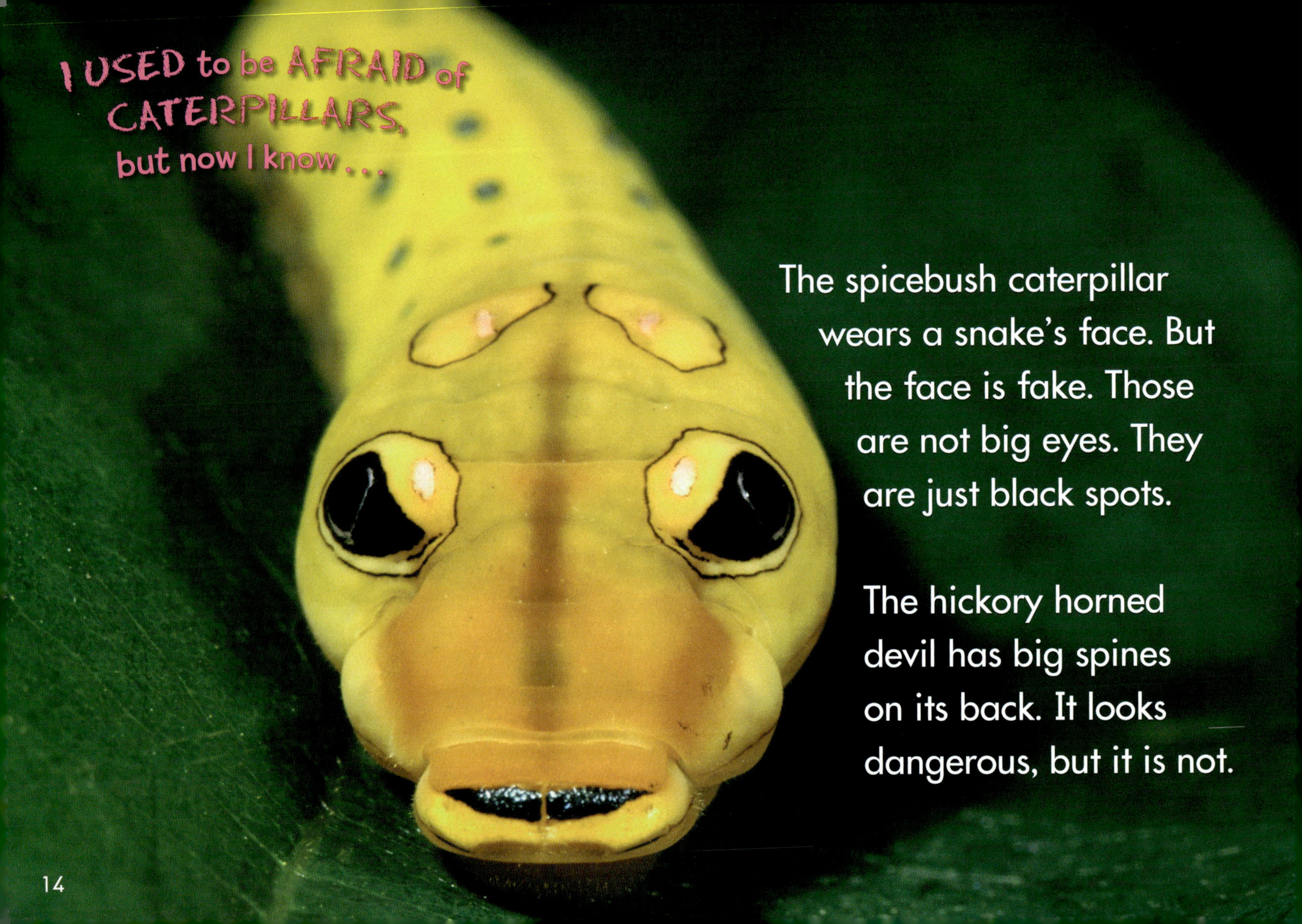

I USED to be AFRAID of CATERPILLARS, but now I know . . .

The spicebush caterpillar wears a snake's face. But the face is fake. Those are not big eyes. They are just black spots.

The hickory horned devil has big spines on its back. It looks dangerous, but it is not.

These caterpillars may scare off hungry birds with their fierce looks. But they will not hurt you.
And now that you know how great caterpillars are, you won't be afraid.

I USED to be AFRAID of CATERPILLARS, but now I know . . .

A few caterpillars have bristles that sting. This green caterpillar is one. These caterpillars will not bother you. It is safe to watch them, but do not touch them.